Know Your Bees

Jack Byard

Old Pond
PUBLISHING

Know Your Bees

Old Pond Publishing is an imprint of Fox Chapel Publishers International Ltd.

First published 2016

ISBN 978-1-910456-12-5

A catalogue record for this book is available from the British Library

Fox Chapel Publishing

903 Square Street Mount Joy, PA 17552

www.oldpond.com

We are always looking for talented authors. To submit an idea, please send a brief inquiry to acquisitions@foxchapelpublishing.com.

Printed in China

Second printing

Book layout by Servis Filmsetting Ltd, Stockport, Cheshire

Picture credits:
(1) Brian Eversham, (2) Tim Melling, (3) Mick Massie, (4) Jaco Visser, (5) Gus Jones, BSCG, (6) Eric Yeomans, (7) Robert Felton Photography, (8, 28) R. C. Knight, (9) Tim Ransom ©, (10) Claire Sell, (11) Ben Smart, (12) Rosemary Winnall, (13) L. Hebdon, (14) Henrique Oliveira Pires, (15) Brian Eversham, (16) Hugh Matthews, (17) Martin Olofsson, (18) Fotoopa, (19) Pietro Niolu, (20, 21) Kate Ferris, (22) Sarah Gregg, (23) Jon Lees, (24) Ashley Perkins, (25) Nigel P. Jones, (26) Andrzej Chojnacki, (27) Caroline Morgan, (29) John Oates, (30) Picture copyright 2004 J.K. Lindsey, (31) Vlad Proklov, (32) Ilona Loser

Acknowledgements

The following are just a few of the many dozens of people and organisations who offered their help, without too much arm twisting, and without whose help and generosity this project would have been impossible.

Susanna Bird of the Gloucestershire Wildlife Trust who from my first bumbling steps into researching the many species pointed me in the right direction.

My gratitude to Darryl Cox and Anthony McCluskey of the Bumblebee Conservation Trust for their advice and patience in answering my countless questions. To Leslie Hebdon, Tim Melling and John Oates, bee experts and enthusiasts, without whose help, advice and generosity the book would have been less colourful and informative.

To my wife Elaine for constantly proofreading, and my granddaughter Rebecca, and friends Sophia and Lauren for their enthusiasm.

Foreword

The honeybee has evolved over 100 million years, and for over 100 thousand years we have been using honey as an antibiotic and sweetener, and now over a third of food crops are pollinated by bees. The bee population has been in decline for over 50 years; sadly, with flower meadows on the decline, this is not a new phenomenon. Food growers are under constant pressure from the major food retailers to produce and sell at unsustainable profit levels. This leads to compromise – every square metre of land must be productive, and so many of the flower meadows have disappeared and field borders have been reduced to the minimum. Now the bee is taking the hit as suitable feeding grounds are reduced or disappear altogether.

Bees also face problems with the Varroa mite, which first appeared in 1992. The mite attaches itself to the back of the bee and slowly over a period of time drinks its blood until eventually the bee dies. Over this period, the weakened bee does not pollinate or feed efficiently, becomes disoriented, and fails to return to its hive. It is believed that this is a major cause of CCD, Colony Collapse Disorder, when the hives break down and fail to do their job and the colony dies.

No single person, society or political organisation has all the answers or is to blame for the current situation; finger pointing and arguing will not solve the problem. We all have a responsibility to protect our bees. The honeybee and bumblebee work for nothing and only ask for food and shelter in return. Overall they are peaceful, non-aggressive little creatures that only attack and sting when they feel threatened. So if a bee is buzzing near you, annoying though it is, the best thing you can do is remain calm and let it fly away in its own time, or gently blow it away rather than flapping about, which is the easiest way to cause you and the bee harm.

I read that in Native American symbolism the bumblebee represents honesty, pure thinking, willingness and drive. What a beautiful symbol.

Contents

Barbut's Cuckoo Bee

(Bombus barbutellus)

Name: *Bombus barbutellus*
Location: Europe
Size: 15–18mm
Habitat: Bramble, knapweed,
lavender and
honeysuckle

The Barbut's Cuckoo Bee is a common sight in Europe, from the north of Spain to the Russian border. In the British Isles it is mainly found in the south of the country, East Anglia and out to the western reaches of Wales. The Barbut's is a cuckoo bee, a squatter, taking over someone else's home for their own purpose and then leaving. The BCB female usually takes over the nest of the Garden Bumblebee, kills the incumbent queen, lays her eggs and leaves it to the Garden Bumblebee workers to bring up its brood. The BCBs are the wastrels of the bee world; they never produce workers.

Both male and female have a broad yellow scarf and a fringe of yellow hair below the chest; the body may have a thin band of yellow hair and a white tail. Female 18mm (.71"), male 15mm (.59") average.

Bilberry Bumblebee

(Bombus monticola)

Name: *Bombus monticola*
Location: Western Europe
Size: 12–16mm
Habitat: Woodlands/moorlands
Population: In decline

The Bilberry Bumblebee or Mountain Bumblebee can be found throughout Europe and in the British Isles on the upper moorlands, grasslands and mountains. In Ireland it is found only in County Antrim, Dublin, Wicklow, Carlow and Wexford. Their small nests are built in the plant life around heather and blaeberry plants (bilberry) and seldom hold more than 100 bees. The queens are on the wing from April to May. The workers do not appear before May and fly until August when they can be seen visiting the flowers of the blueberry, heather, white clover and bird's-foot trefoil. Once a common sight but, regrettably, BB numbers are now in decline.

The Bilberry Bumblebee is smaller than other bumblebees. The queen and males are very similar in appearance – the queens, males and workers have a broad band around the chest with a narrow band at the rear, half of the body is ginger-red and the male has yellow hairs on the face. Size: queen 16mm (.63"), male and workers 14mm (.55"), drones 12mm (.42").

Blue Mason Bee

(Osmia caerulescens)

Name: *Osmia caerulescens*
Location: Western Europe
Size: 8–10mm
Habitat: Woodlands, forests,
flower meadows,
gardens and orchards

This tiny bee is found throughout Europe and most parts of England and Wales – it is even found high in the Alps. The BM is frequently mistaken for a fly. It is only its hairy legs that gives it away – I know the feeling! The nests are usually made and lined with chewed leaves and flower petals; the nesting area is widespread from quarries to holes in dead tree stumps and branches, dry stone walls, cliffs, thatched roofs and brambles. It can be seen flying from March until August at the edges of woodlands and forests, pollinating flower meadows, garden herbs, peas, beans and orchards.

The female's body is blue/black with a metallic lustre; the males are bronze with pale yellow hair, a brassy green face and green eyes.
Size: female 9–10mm (.35–.39"), males and workers 8–9mm (.31–.35").

Brassy Mining Bee

(Lasioglossum morio)

Name: *Lasioglossum morio*
Location: Western Europe
Size: 5–8mm
Habitat: Exposed soil/sloping
soil banks

This tiny bee is frequently mistaken for a fly and it can be found buzzing around in most of Europe and the UK, from the Isles of Scilly through to Wales and up to Cumbria. They have as yet not ventured into Scotland or Ireland – they do not know what they are missing! They build their nests in exposed soil and sloping soil banks and in the cavities of old walls. They can be seen on the wing from April to July in gardens, open countryside, parks and roadside verges. The type of flowers they visit is large and diverse; they include the alder buckthorn, burdock, dandelion, bramble and willow.

Slim and black with clear bright bands around the body. The head and body are metallic bronzy green. Size: 5–8mm (.19–.31").

Broken-Belted Bumblebee

(Bombus soroeensis)

Name: *Bombus soroeensis*
Location: Western Europe
Size: 12–16mm
Habitat: Abandoned burrows and nests
Population: Under threat of extinction

The Broken-Belted Bumblebee is found pollinating and feeding throughout Europe. Although listed as being seen over most of the British Isles, it is more readily seen in the north and Scotland including Skye, but not on the outer islands. The preferred nesting sites are abandoned burrows and nests of mice and voles. The colony sizes are quoted as housing from 80 to 150 bees, but seldom reach 100. They can be seen flying from June to September over heaths and moorlands and chalky grasslands, feeding and pollinating peas, beans, bell flowers and devil's-bit scabious. The last census a number of years ago agreed that the Broken-Belted Bumblebee was not under threat of extinction. It is now suggested that another look should be taken.

All the staff have two lemon-yellow stripes, one on the chest and one lower down. This one is often split, but not always (see image). Size: queen 16mm (.63"), workers 12mm (.47"), males 13mm (.51") average.

Brown Banded Carder

(Bombus humilis)

Name: *Bombus humilis*
Location: Europe, Tibetan plateau, northern China
Size: 10–18mm
Habitat: Tufts of grass

The Brown Banded Carder is, in the British Isles, a rare beastie. The BBC is a common sight in Europe and as far afield as the Tibetan plateau and in northern China. In these fair isles, however, it is only found on the chalk landscapes and coasts of southern Britain. Even though Britain is at the very edge of its territory, it was once widespread, feeding on the flowers in rich open meadows. These meadows are now in serious decline, mainly through intensive farming, and the BBC is unable to adapt to the changing conditions. Its small nests are usually found in tufts of grass and the BBC can be seen flying from May to September. The size of colonies seldom reaches three figures.

The body is tawny coloured with beige sides and a broad brown band on the upper body.
Size: queen 16–18mm (.63–.71") male and workers 10–15mm (.39–.59").

Buff-Tailed Bumblebee

(Bombus terresiris)

Name: *Bombus terresiris*
Location: Europe, North African coast, central Asia
Size: 10–22mm
Habitat: Abandoned mouse holes
Population: Most numerous of the bumblebees in Europe

The Buff-Tailed Bumblebee is also known as the Large Earth Bumblebee and is the most numerous of the bumblebees in Europe and usually the first to be seen in spring, sometimes as early as February. It is also at home on the North African coast and in west and central Asia, and like many other BBs nests in abandoned mouse holes. The colonies normally number about 150 bees. Since the mid 1980s, they have been bred commercially in Europe where they are mainly used for pollinating tomatoes in greenhouses, a task that used to be done by hand. In 2008 the live import of the BTB was banned by the Australian government, fearing it would create havoc amongst the native flora and fauna. It is also classed as an invasive alien species in Japan.

The queen has dirty-orange-coloured hairs at the end of the body; the workers have a white tail. The queen averages 20—22mm (.79—.87"), males and workers 10—16mm (.39—.63").

Common Carder

Name: *Bombus pascorum*
Location: British Isles
Size: 13mm
Habitat: Old mouse nests

(Bombus pascorum)

The Common Carder is a bumblebee. The queen searches for a nesting site, looking for small holes, both above or below the ground. She prefers nesting in old mouse nests, but bird nests, small cavities inside barns and even your garden shed will be deemed acceptable. The queen will fill the cavity with moss and grass, a small area will be filled with pollen to feed the young and another area will have a store of nectar to feed them in the bad weather when they are unable to fly. They can be seen throughout the British Isles from March to November feeding on many varieties of wild flowers and fruit trees. The transfer of pollen from plant to plant is an essential part of our ecosystem for the pollination of our horticultural crops.

Brown and orange shaggy hair; occasionally has brown bands on body. Length 13mm (0.51") average.

Davies Mining Bee

(Colletes daviesanus)

Name: *Colletes daviesanus*
Location: Central and Northern Europe
Size: 7–9mm
Habitat: Sand and gravel pits, sandstone cliffs and clay banks

The Davies Mining Bee is one of the nine species of solitary mining bee in the British Isles, and one of the estimated 700 species worldwide. The Davies Mining Bee is found throughout central Europe and north to Sweden and Finland. They can be seen flitting their stuff through England, Wales and the Channel Islands but they are a little scarce in the flower gardens of Scotland and Ireland. They drill 5mm-ish holes in sand and gravel pits, sandstone cliffs and clay banks, even in soft mortar in brickwork, and line the nests with a plastic, polyester secretion, making the nests waterproof. They are a common sight from June to August and can be seen flying, feeding and pollinating flower gardens and meadows. Daisies, ragwort and asters are a favourite.

The head and chest are a hairy reddish brown; the body is black with a cover of light grey hair and a pointed tail. Size: 7–9mm (.28–.35") average.

Early Bumblebee

Name: *Bombus pratorum*
Location: Western Europe
Size: 10–17mm
Habitat: Old birds' nests

(Bombus pratorum)

The Early Bumblebee is a common sight in most of the British Isles and Europe, though not often seen in North West Scotland. The EB flies early and can be seen in the southern parts of the British Isles in February. They will build a colony in an old bird's nest, a hole in your garden rockery or an abandoned mouse nest; the colonies seldom exceed 100 bees.

They are usually found on lavender, thistle, sage, raspberries and dandelion and many daisy-like flowers, just a small sample of the flora that the small EB feeds on and pollinates. All bees are important in the pollinating of our crops.

The queen is black with a yellow collar with another yellow band around the body and a red tail; the male has two wider yellow bands on the body and a red tail.
Size: queen 15–17mm (.59–.67"), males and workers 10–14mm (.39–.55").

Early Mining Bee

(Andrena haemorrhoa)

Name: *Andrena haemorrhoa*
Location: Worldwide
Size: 8–11mm
Habitat: Sandy, loamy soil

The Early Mining Bee is a member of the large Andrena genus of mining or sand bees; there are 1300 different species worldwide and almost 60 living in the British Isles. They prefer to dig their nests in sandy, loamy soil. At the bottom of each shaft they lay an egg on top of a ball of pollen and nectar, then the nest is sealed and the egg left to develop. They can be seen flying from March to June, feeding and pollinating a wide range of crops from dandelion to elder and daisies. They are also seen in oil-seed rape areas, where despite disturbance to the soil when planting, the nests remain intact, allowing the Early Mining Bee to thrive and pollinate the crop.

The body is a shiny black; the female has a rusty-red chest and a rusty-red tip to the tail. The male is much smaller and lighter in colour within the colour spectrum, tipping more to grey or white.
Size: 8–11mm (.31–.43") average.

European Honeybee

Name: *Apis mellifera*
Location: Worldwide
Size: 10–20mm
Habitat: Gardens, woodlands and
flower meadows

(Apis mellifera)

The European Honeybee originated in Africa and has spread throughout Europe and the world. It was introduced into America in the early 17th century and has since spread throughout the Americas and as far north as Alaska, providing honey and pollinating foodstuffs that we require to stay alive, along the way. Most of us think of the honeybee living in managed wooden hives, but there is a wild variety that nests in hollow trees. A colony of the EHB can be from 30,000 to 80,000 individuals, with the queen laying up to 2000 eggs daily and living for about 3 years. They forage for food, flying up to 6.5km (4 miles) from the hive, in gardens, open woodlands, flower meadows and agricultural land, providing us with honey.

They vary in colour but mainly have burnt-orange and darker-black alternating stripes. The European Honeybee has 4 wings, 6 legs, large eyes and a pointed body.
Size: queen 18–20mm (.71–.79"), males and workers 10–15mm (.39–.59").

Fabricius' Nomad Bee

(*Nomada fabriciana*)

Name: *Nomada fabriciana*
Location: Europe
Size: 8–10mm
Habitat: Chalk/limestone
          grasslands

The Fabricius' Nomad Bee is found throughout Europe; in the UK it is to be seen mainly in the south of the country and Wales, Rutland and Leicestershire and very infrequently in Scotland and Ireland. The FNB are cuckoo bees: they sneak into the nests of the Gwynne's Mining Bee, lay their eggs and leave. They are also known to take over the nests by force, killing the female in the colony. The FNB are bivoltine, meaning they create two generations a year. The first generation can be seen flying over chalk and limestone grasslands from March to June and the second generation June to August, sometimes overlapping. Areas where abandoned vineyards, dandelion, willow and rape are found are their favourite dining areas.

The females have red and black banded antennae. Both males and females have black and reddish bands around the body; the female can also have yellow spots.
Size: queen 8–10mm (.31–.39").

Forest Cuckoo Bumblebee

(Bombus sylvestris)

Name: *Bombus sylvestris*
Location: Britain
Size: 14–15mm
Habitat: Lavender, white deadnettle, bramble and buttercup flower gardens

The Forest Cuckoo Bumblebee is found all over Britain and, as the name suggests, is the cuckoo of the bee world, laying its eggs in the nests of other bumblebees. The nests preferred are those of the Early BB, Heath BB and the Bilberry BB. The FCB lays her eggs in the host's nest and leaves it up to the workers to rear her young. They do not forage with any great effort, and they have only themselves to look after, so they spend their days lazing around drinking nectar from lavender, white deadnettle, bramble and buttercup, to name but a few.

Usually the male and female have a yellow stripe on the upper body and another lower down. The tail is white with a black tip for the female; the tip of the male is ginger. In Scotland the body may be black with a yellow band to the tail.
Size: queen 15mm (.59"), male 14mm (.55").

Garden Bumblebee

Name: *Bombus hortorum*
Location: British Isles
Size: 11–20mm
Habitat: Grassy banks/tree roots

(Bombus hortorum)

The Garden Bumblebee has the longest tongue of all British bumblebees: it is the same length as its body, and despite the large number of BB species within these islands it is one of the few remaining with a long tongue. The GB usually builds its nest underground, but also builds in grassy banks and tree roots – even in the bird box at the bottom of your garden. The size of a GB colony seldom exceeds 100 bees. They can be seen flying from March to October, feeding mainly on deep tubular flowers such as foxglove, daffodil, honeysuckle and bluebell, to name but a few. The GB can be seen all over the British Isles and as far north as Orkney and Shetland.

The queen is large and scruffy looking with long black hair with two yellow bands at the upper part of the body and one at the lower end plus a white tail. Size: males 17–20mm (.67–.79") and workers 11–16mm (.43–.63").

Great Yellow Bumblebee

(Bombus distinguendus)

Name: *Bombus distinguendus*
Location: Europe
Size: 12–21mm
Habitat: Low-lying grass plains/ machair
Population: Dropped by 80% over the last 100 years

The Great Yellow Bumblebee is found throughout Europe. Once a reasonably common sight in the UK, it is now clinging on to life with its wing tips in the outer reaches of Scotland, Arran, Burren and the Mullet peninsula and lowland Caithness. The GYB requires low-lying grass plains, machair – with its abundance of flowers and red clover – its favourite, knapweed, pea, thistle and mint. This type of habitat has been in decline since the 1940s when red clover ceased to be grown, meaning the population of the GYB has dropped by 80% over the last 100 years. In 2010 11 farms around Caithness were sown with the GYB's favourite plants in an attempt to reverse the declining numbers. The nests are built in abandoned mouse holes, sand dunes and grass tussocks. The GYB can be seen flying from May to September.

Mainly lemon yellow/
straw coloured, with a
black band intermixed
with yellow hairs, at
the base of wings.
Size: 12—21mm
(.47—.83") average.

Hairy Footed Flower Bee

(*Anthophora plumipes*)

Name: *Anthophora plumipes*
Location: Europe, East/North Africa, USA
Size: 15–17mm
Habitat: Clay slopes/mud walls

The Hairy Footed Flower Bee can be found in southern and central England and Wales and as yet has not been found any further north or in Ireland. They can be found in most of Europe, the Near East and North Africa and were only introduced into the USA in the 20th century. They usually nest in clay slopes and steep mud walls. If February is warm they will emerge and fly until the middle of June. They have a swift darting flight and can be seen hovering and darting around their favourite plants, the blue and pink lungwort, comfrey and deadnettle, with their 14mm tongue extended ready for the delicious nectar.

The female is black and furry with red and orange hair on the back, resembling a bumblebee, which it is not. The males are a rusty brown with long hairs on the legs, hence the name.
Size: 15–17mm (.59–.67").

Heath Bumblebee

(Bombus Jonellus)

Name: *Bombus Jonellus*
Location: Europe, Canada, Alaska
Size: 12–16mm
Habitat: Gardens, moorlands and
heaths

The Heath Bumblebee is found throughout Europe, Iceland, Scandinavia and Canada to the Hudson Bay and Alaska. In the UK, though widespread, the distribution is rather erratic, from East Anglia through to Scotland, the Orkneys, the Hebrides and the Shetland Isles but rarely seen on the southern east coast. They build their nests in old vole and mouse holes, abandoned birds' nests and roof voids, with a colony average of about 50. The Forest Cuckoo Bee is a parasite of the HBB. Forget the name: they are found in gardens, parks and moorlands as well as heaths, feeding on clover, thistle and cowberry.

All the staff – queens, males and the workers – have black faces, sometimes with a yellow patch on top. All have three yellow bands around the body – one around the chest and two around the body – and a white tail. Those in the Western Isles and Shetland have a red tail.
Size: queen 16mm (.63"), the males and workers 12mm (.47").

Ivy Mining Bee

Name: *Colletes hederae*
Location: Europe
Size: 10–14mm
Habitat: Loose sandy soil

(Colletes hederae)

The Ivy Mining Bee prefers to build its nest in loose sandy soil, on south-facing banks that have a minimum of vegetation, burrowing up to 30cm (12" approx) into the bank. It is classed as a mining bee. In an ideal nest-building area there can be thousands of nests. The Ivy Mining Bee lines its nests with a plastic-like material to make it waterproof, hence its other name: the plasterer bee. It is found in most of Europe but did not appear on these shores until 2001. In the UK it is found mainly in the Channel Islands, southern England, Norfolk, Staffordshire, Shropshire and south Wales. It is a late flyer, emerging in September, coinciding with the flowering of the ivy, its main food source, and if the weather is warm, is seen until November.

The upper body is covered with orange-brown hair, and the lower body has several orange-brown hair bands.
Size: female 13–14mm (.51–55"), males and workers 10–11mm (.39–.43").

Large Garden Bumblebee

(Bombus ruderatus)

Name: *Bombus ruderatus*
Location: Europe
Size: 11–23mm
Habitat: Flower-rich meadows
Population: Species at risk

The Large Garden Bumblebee is found throughout central Europe but not too often seen in Spain. In the British Isles it is now mainly centred along the river valleys in the central and southern regions but has not been seen flying and pollinating in Wales since the middle of the 1960s. In the north of the country it has been sighted in Northumberland, but nary a one in Scotland. It builds its nest in abandoned vole and mouse burrows. The colonies usually have between 50 to 100 inhabitants, and they can be seen buzzing around from April to August. The LGB prefers to feed in sumptuous flower-rich meadows, but these are increasingly difficult to find. This at-risk species feeds on red clover and foxglove. In preservation planted areas it has taken a liking to legumes and comfrey.

Overall black with two yellow bands on the chest and a single yellow band on the body. The tail is white. Size: queen 21–23mm (.83–.90"), males 15–16mm (.59–.63") and workers 11–18mm (.43–.71").

Moss Carder Bee

(Bombus muscorum)

Name: *Bombus muscorum*
Location: Mongolia, Scandinavia, Russia
Size: 14—19mm
Habitat: Salt marshes, sandy hills and boggy ground
Population: In decline

Not a common sight in Europe but you will see them on your next holiday in Mongolia, Scandinavia, parts of Russia and Crete. In the British Isles the populations are patchy to say the least, and are most likely spotted on the northern coasts of Scotland, the Isle of Lewis, the Hebrides, Orkney and Shetland, plus there are small populations on the Isle of Man and in Ireland. They build their nests on salt marshes, sandy hills and boggy ground. They will occasionally build them just under the surface but more usually at ground level covered with dry moss and grass. These colonies seldom exceed 100. The MCB is in flight from May to August feeding on clover, thistle, bird's-foot trefoil and vetches. The loss of flowering grassland is the major cause of their decline.

The upper body is bright orange and the lower body is yellow with black hairs. The face, legs and under body of the Hebridean, pictured, are black. Size: queen 19mm (.75"), males and workers 14mm (.55").

Red Mason Bee

Name: *Osmia rufa and Osmia bicornis*
Location: Europe, Georgia, Iran
Size: 7–14mm
Habitat: Local parks, allotments and gardens

(Osmia rufa and Osmia bicornis)

The Red Mason Bee can be found throughout Europe, Norway and Sweden, out to Georgia and into Iran. It can be seen in most parts of England and Wales and as far north as Perthshire in Scotland but maybe not quite as often. The nests are built in the crevices of walls, clay banks and hollow plant stems, and the RMB reuses holes that have been made by other insects, or even a discarded snail shell. The nests are sealed by 'plastering' over the entrance to the nest with mud; this is where they get the 'mason' in their name. They can be seen on the wing from late March to the end of July, pollinating and feeding from their favourite food source, apple and pear trees, but also visiting local parks, allotments and gardens. They are now being bred commercially as an efficient, environmentally friendly pollinating machine.

The bodies of the male
and female are covered
with thick ginger
hair; the male, who is
smaller, has a tuft of
white hair on its face.
Size: 7–14mm
(.28–.55").

Red-Shanked Carder Bee

(Bombus ruderarius)

Name: *Bombus ruderarius*
Location: EU, North West China,
Siberia, North Africa
Size: 13–17mm
Habitat: Flower meadows and
open grasslands
Population: Believed to be in
decline

Found throughout the EU, North West China, Siberia and North Africa. In the British Isles it is found mainly in the south of the country and you will not see it again until you reach the west of Scotland. The nests are usually built in abandoned mouse and vole nests and are lined with moss and grass. A colony consists of 50–100 bees. The RSCB can be seen flying from mid April to early September, flying over flower meadows and open grassland, feeding off the flowers of peas, beans, deadnettle and clover. In 1991 it was believed the species was not at risk, but since then numbers have fallen dramatically. The highest-populated areas are where there is unimproved grassland and non-intensive farming. The entire bee population helps to pollinate our crops, which in turn help feed us. We cannot allow any species to become extinct.

Mainly black with a red tail. Simple.
Size: queen 17mm (.67"), male 13mm (.51"), worker 15mm (.59") average.

Red-Tailed Bumblebee

(Bombus lapidarius)

Name: *Bombus lapidarius*
Location: Western Europe
Size: 11–22mm
Habitat: Hedgerows, gardens and
farmland
Population: In decline

The Red-Tailed Bumblebee is one of the most common and easily recognised of the bumblebee species. It is widely spread throughout England but not seen as readily in Scotland, Wales and Northern Ireland. Despite being common, the numbers of the RTB are slowly dropping, mainly due to the decline in their food sources. Their nests can be found under stones and in dry stone walls and usually contain a maximum of 50 eggs. The RTB can be found flitting around the countryside from April to September. Their preferred food is the nectar of clover and deadnettle, but they can be found in hedgerows, gardens and farmland – anywhere they can find flowers to feed on.

The queen has a large, round, black, hairy body with a big red tail up to 20mm long. The smaller males, sometimes no larger than a housefly, have a yellow band on the body and a yellow face and red tail.
Size: queen 20–22mm (.79–.87"), males and workers 11–16mm (.43–.63").

Sharp-Tailed Bee

Name: *Coelioxys inermis*
Location: British Isles
Size: 10–13mm
Habitat: Daisy-type flower
gardens

(Coelioxys inermis)

The Sharp-Tailed Bee is widespread throughout the British Isles from the south to Cumbria in the north and can live at altitudes of up to 1486m (4875'). They fly from late spring through the summer, feeding off daisy-type flowers, but not a great deal seems to be known about them and sightings are very rare. One bee expert I spoke to, who has studied bees for 25 years, has only encountered the Sharp-Tailed Bee about 20 times in that period. The Sharp-Tailed Bee is a cuckoo bee, using the nest of the Willughby's Leafcutter Bee. The sharp tail is used to pierce holes in the leaf-lined cells, opening up the cells for the female to lay her eggs.

A long, tapered, triangular body with a sharp tail, and the face has golden and white hairs and hairy eyes. Size: 10–13mm (.39–.51").

Shrill Carder Bee

(Bombus sylvarium)

Name: *Bombus sylvarium*
Location: Europe
Size: 10–18mm
Habitat: Flower- and herb-rich
grasslands
Population: In decline

The Shrill Carder is one of the smallest of the bumblebees. It is a common sight throughout Europe and was once so in the British Isles, in the 1800s and the early part of the last century, until the decline started in the 1970s. It is is now classed as endangered, however, and only found in Somerset, Essex, Kent, Wiltshire and South Wales, military ranges and unimproved pastures. Survival depends on flower- and herb-rich grasslands that have not been treated with pesticides and herbicides. Abandoned mouse and vole nests are their preferred home, each nest containing in the region of 100 bees. They fly from April to September, the queen making a high-pitched shrill buzz. The males and workers are less noisy, of course.

Grey-green in colour with a single black band across the upper body and two dark bands on the lower body.
Size: queen 16–18mm (.63–.71"), males and workers 10–15mm (.39–.59").

Southern Cuckoo Bumblebee

(Bombus vestalis)

Name: *Bombus vestalis*
Location: Europe, Western Asia
Size: 15–22mm
Habitat: White clover knapweed, blackthorn, dandelion and sallow

The Southern Cuckoo Bumblebee is found in most of Europe and out to western Asia. In the British Isles it was thought until quite recently not to travel further north than the river Humber, but in the summer of 2009 it was spotted in southern Scotland. The SCB uses the Buff-Tailed Bumblebee as a host. The queen forces her way into the nest, kills the resident queen, lays her eggs and leaves her brood to be raised by the Buff-Tailed workers. Cuckoo bees have no need to forage and store food: they eat and drink when they feel like it, having only themselves to look after. They can be seen flying from April to July on their favourites: white clover knapweed, blackthorn, dandelion and sallow.

Mostly black with an orange collar, and the middle of the body has a yellow band. The tail is mostly white. Size: queen 20–22mm (.79–.87"), males 15–17mm (.59–.67").

Tawny Mining Bee

(*Andrena fulva*)

Name: *Andrena fulva*
Location: Central Europe,
British Isles
Size: 8–12mm
Habitat: Fruit trees, garden
plants

Seen throughout central Europe and most of the British Isles, though a little thin on the ground in Scotland. They build nests in well-mown lawns, grassy banks and flower beds. The female digs a vertical shaft approximately 4mm (.16") in diameter and up to 300mm (12") deep; she will then dig branches off this shaft. Each branch is filled with pollen and nectar into which she lays one egg. The only sign the bees are living in your garden are tiny volcano-shaped mounds of earth. There can be several individual nests within a small area, although the Tawny does not damage your lawn. Please do not try to move them: they are harmless, non-aggressive little fellows and will soon be gone. They fly from April to June, feeding and pollinating fruit trees, your garden plants and gooseberries.

The females have ginger/red hair on the back, the males are slightly more yellow and slimmer, and both have black under bodies.
Size: female 8–10mm (.31–.39"), males and workers 10–12mm (.39–.47").

Tree Bumblebee

(*Bombus hypnorum*)

Name: *Bombus hypnorum*
Location: Europe
Size: 10–mm
Habitat: Well-mown lawns,
grassy banks and
flower beds

The Tree Bumblebee can be found across Europe, but only arrived in the British Isles at the beginning of the 21st century, and in the space of 15 years has expanded across most of England, Wales and southern Scotland. As the newcomer on the block, there were concerns as to how it would interact or affect the local inhabitants – bees, that is – but these fears have proved foundless. They tend to build nests in trees or tall buildings and under roof tiles, but compost heaps and even bird boxes are used. A colony consists of anywhere from 150 to 400 bees. They can be found from March to the late summer, feeding on a wide range of plants, from rhododendron to raspberries, gooseberries, cotoneaster and strawberries. The Tree Bumblebee is a first-class pollinator.

The upper body is ginger; the rest of the body is covered with black hair, apart from the tail, which is always white.
Size: queen 22mm (.87"), males and workers 10–16mm (.39–.63").

White-Tailed Bumblebee

(Bombus lucorum)

Name: *Bombus lucorum*
Location: Western Europe
Size: 11–17mm
Habitat: Gardens, farmland,
hedgerows, moorlands
and heathlands

White-Tailed Bumblebees build their nests in abandoned mouse and vole nests. The colonies usually consist of 100 to 300 bees, but colonies of 400 are not unknown. The WTB is an early riser and can be seen flitting about the countryside from February to November. It can be found in gardens, farmland, hedgerows, moorlands and heathlands throughout the British Isles and Europe. In early spring you will find them hovering around fruit trees. The WTB has, for a bumblebee, a very short tongue and tends to forage on flowers similar to daisies; some have a sneaky habit of cutting a hole in the base of the flower so they can drink the nectar.

A black hairy body with a lemon-yellow band on the second section of the body and a white tail. The males have much thicker yellow hair and have a bushy appearance. Size: queen 14–16mm (.55–.63"), male and worker 11–17mm (.43–.67").

Willughby's Leafcutter Bee

(*Mesachie willughbiella*)

Name: *Mesachie willughbiella*
Location: Europe
Size: 15–17mm
Habitat: Woody plant stems, ground, and soil of flower pots

This leafcutter bee is named after the naturalist Frances Willoughby (Willughby) (1635–1672). The WLB is found throughout Europe and in most areas of the UK, but in lesser numbers in the north of the country and Scotland. Their nests are built in warm sheltered locations, in woody plant stems, preferably willow, but also in the ground or the soil of flower pots. They fly from June to August and can be seen carrying semi-circular sections of leaves, which are used to build their nests. Their leaf of choice is from the rose bush, but aster, dandelion, thistle and bellflower are on their wish list. The WLB is only one of over 1500 species of leafcutter bee.

The female is a hairy yellowish brown; the hairs on the pollen sac underneath the body are bright orange. The males have the same overall body colour but with a whitish under body and the ends of the legs have whitish hairs.
Size: 15–17mm (.59–.67").

Wool Carder Bee

(Anthidium maicatum)

Name: *Anthidium maicatum*
Location: Europe, North Africa,
  Asia, the Americas
Size: 11–17mm
Habitat: Hollow stems and
  cavities

This large solid-looking bee is one of the largest in Britain and is found mainly on the south coast, Wales, the Isles of Scilly and the Channel Islands, with small numbers in the north of England. It has been sighted in Edinburgh, Dumfries and Galloway and is found throughout Europe, North Africa, Asia, North America and various sites in South America. The WCB arrived in the USA by accident in the early 1960s and like all good bees has spread, and is now found across America and South East Canada. The Wool Carder can be seen from late May until the first weeks in August, feeding on lavender, marjoram, thyme and rosemary, chasing off any insects that invade its territory.

Mainly black with yellow spots; pale hair on upper body and yellow markings on the face and sides of the body. The male has 5 spikes on the tail. Size: female 11–13mm (.43–.51"), males 14–17mm (.55–.67").

Also in the 'Know Your' series...

Know Your Goats

An illustrated guide to the identification of goats likely to be encountered in Britain today.

Know Your Sheep

Forty-one breeds of sheep that can be seen on Britain's farms today.

Know Your Dogs

Forty–five popular dog breeds, from Dalmatians to Corgis, each accompanied with text describing their history, characteristics and abilities

Know Your Cats

Covers forty-two breeds, each accompanied by a clear
description covering the history, appearance and personality
of the breed.

Know Your Donkeys

An enchanting sample of thirty-four
breeds of donkeys and mules from
around the world, from the miniature
to the mammoth.

Know Your Ducks

Some of these forty-four breeds of duck are native to Britain,
whilst others simply choose to holiday here. Each breed's flaying
and egg-laying abilities are noted.

To order any of these titles please contact us at:
Tel: 0114 240 9930 • Email: contact@oldpond.com • www.oldpond.com